# Fuelling the Future

**Richard Hatton**
Photographs by Jenny Matthews
Illustrations by Peter Bull Art Studio

A & C Black · London

333

# Contents

**Cover photographs**
Front – Electrical energy test (see page 18)
Back – Power station and pylons (see page 16)

**Title page photograph** – Windmill and wind farm
(see page 27)

Acknowledgements
Photographs by Jenny Matthews except for: p.4 Ed
Barber; p.9 Mark Edwards, Still Pictures; p.10 British
Coal Opencast; pp.11, 12, 13, 14 (bottom) Mark
Edwards, Still Pictures; p.16 Centre for Alternative
Technology; p.17 Mark Edwards, Still Pictures;
pp.22–23 British Nuclear Fuels, plc; p.24 Centre for
Alternative Technology; pp.25 (bottom), 26, 27, 28
(top), 29 Mark Edwards, Still Pictures; p.30 GSF
Picture Library, W. Higgs.

Illustrations by Peter Bull Art Studio

The author and publisher would like to thank the
following people for their invaluable help during the
preparation of this book: the staff and pupils of
Grasmere J.M.I. School and Berger J.M. School.

A CIP record for this book is available from the
British Library.

ISBN 0-7136-3540-1

First published 1992 A & C Black (Publishers) Ltd
35 Bedford Row, London WC1R 4JH

© 1992 A & C Black (Publishers) Ltd

Typeset by Rowland Phototypesetting Ltd,
Bury St Edmunds, Suffolk
Printed in Italy by Imago

# What is energy?

What do you have in common with a television set, a car, a gas fire and a light bulb? The answer is that you and all these things need energy in order to work. Food, electricity, petrol and gas are some of the things that provide this energy. They are called energy sources.

▼ Fairground machines need energy in order to work and they produce energy too, in the form of light, sound and movement.

The Earth stores energy in many forms such as coal, oil, gas and water. People have only just begun to realise that many of these stores of energy cannot last for ever. Every time you switch on a light, turn on a heater or you're driven somewhere in a car, you use up more of this energy.

As the Earth's population increases each year, more energy is used. What can we do *now* to make sure that there will be enough energy left to fuel the future?

## Your daily energy use

Look through some old magazines and newspapers. Find advertisements for energy sources such as gas and electricity and for objects like cars and electrical gadgets which use energy.

You could cut them out and make them into a poster to show the different types of energy you use every day. Do you think the advertisements encourage people to use more energy than they need?

5

# Energy and you

Your body needs energy to make it work. Walking, running, breathing, keeping warm and even thinking are all activities which use energy. Which do you think uses the most?

Food is a source of energy. It contains chemical energy which is stored in your body. Foods called carbohydrates and fats provide you with most of the energy your body needs.

Carbohydrates such as bread, potatoes, rice and sugar give you energy quickly. Butter, margarine and cooking oil all contain fats which stay in your body for longer and are used up more slowly.

Your muscles use energy from food to make your body move. When you are still, your muscles use very little energy. The faster you make them work, the more energy they need. When you run fast, how does your body tell you when it hasn't got enough energy left?

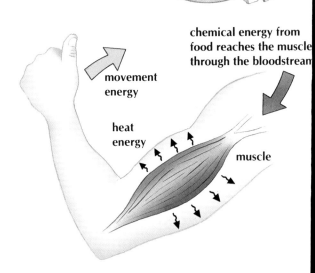

chemical energy from food reaches the muscle through the bloodstream

movement energy

heat energy

muscle

You have some special muscles in your body which, to keep you alive, must never stop moving. They need a continuous supply of energy. Can you think which muscles these are?

It is very important to eat enough food each day to keep your body working properly. Without enough energy from food you would gradually become very weak and die.

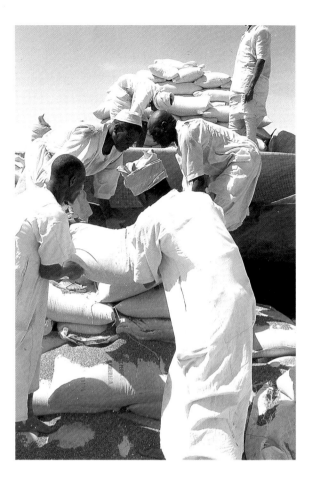

◀ When an area is hit by famine, relief food supplies are urgently needed to provide people with a vital source of energy. As well as emergency food supplies, the people of famine-hit areas need long-term aid in the form of education and equipment which will help them to produce enough food to avoid future outbreaks of famine.

## Energy levels in food

Collect food packaging such as packets, wrappers and cartons. On each piece of packaging you'll find information about how much energy there is in each 100 grams of the food inside. Energy is measured in kilojoules. Which types of food contain the most energy?

# Plants and energy

Where do plants get their energy from? Put a plant on a window sill and watch what happens to it over a few days. As well as water, plants need light – and light is another form of energy.

The sun is a vital energy source. The leaves of a plant use sunlight to make food in a process called photosynthesis. This process changes light energy into chemical energy. A plant stores this energy as food in its stem or roots. We eat plants and get energy from then

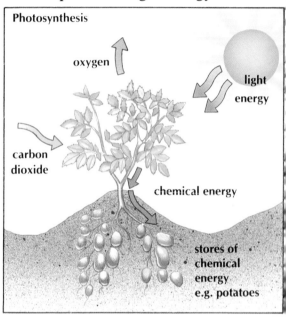

Photosynthesis

oxygen

light energy

carbon dioxide

chemical energy

stores of chemical energy e.g. potatoes

A tree stores chemical energy in the wood of its trunk. We cannot use this energy by eating it, but we can burn the wood to produce heat energy.

People in many countries use wood as a fuel for cooking, heating and driving machinery. But the smoke from burning wood causes air pollution. And all over the world, wood is being used up much faster than new trees can be grown to replace it.

During photosynthesis, plants take in carbon dioxide from the air and give out oxygen. The drop in the number of trees has contributed to a rise in the amount of carbon dioxide in the air. The carbon dioxide traps heat energy from the sun, causing a worldwide rise in temperature. This is called the greenhouse effect.

It is vital that there are enough trees in the world to help to reduce the levels of carbon dioxide in the air.

It's the job of a forester to plant trees that will not damage the local environment or upset the balance of the wildlife in an area. You can help by joining a local tree-planting group and by reducing the need for trees by using less energy.

Scientists monitor the growth of tree seedlings as part of their research into reforestation. The type of seeds that grow best in these tests will be sown directly on to the land to grow into a new forest.

# Earth as an energy source: coal

Wood is an energy source which you can see growing on the surface of the Earth. Hidden beneath the surface there are other sources of energy. Coal is one of these. It is a store of chemical energy formed from the fossilised remains of swampy forests which were covered by the sea 300 million years ago.

Coal is usually buried deep underground, but some is found close to the surface and is dug up in an open pit. If the pit is not filled in and planted with trees after the coal has been mined, it can leave ugly scars on the landscape.

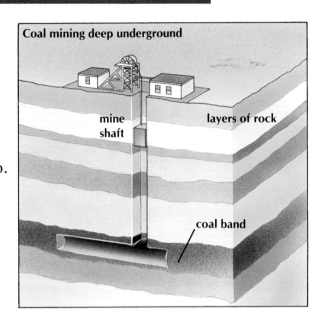

Coal mining deep underground

mine shaft

layers of rock

coal band

◀ The photograph on the left shows a huge spoil heap of the waste material which was produced by opencast mining.

▶ The right-hand photograph shows the same area of land after the spoil heap had been levelled and the ground planted with grass, hedges and trees.

▲ These trees have been damaged by acid rain. The tree trunks are thin and spindly and the branches are almost bare.

Look for evidence of acid rain damage around your local area. Churchyards are a good place to start – sometimes gravestones crumble away in patches where acid rain has dissolved the stone.

When coal is burned it provides more heat energy than wood. But burning coal also produces carbon dioxide, sulphur dioxide and sooty smoke. The carbon dioxide helps to cause the greenhouse effect. And when water in the air mixes with sulphur dioxide gas, acid rain is formed. This gradually kills trees and poisons lakes. Sooty smoke blackens buildings and pollutes the air.

You may not have a coal fire at home, but you still use a lot of energy that has come from coal. Coal is used as a fuel in many power stations where electricity is generated. Every object made of iron or steel in your home or school was produced using energy from coal.

Each year our use of coal increases and now the supplies of coal are running out. Unlike wood, once coal is used it cannot be replaced. In 300 years' time, there will be no coal left. The only way to make the supplies last longer is to use less energy.

# Earth as energy source: oil and gas

Oil and natural gas are energy sources which lie hidden deep beneath the ground. They are formed from the remains of tiny sea creatures that died millions of years ago. Over time, they were crushed beneath a huge weight of mud and rock until gradually they changed into oil and gas.

To reach these energy sources, people drill deep holes into the layers of rock which are covering the reserves of oil and gas. This often has to be done in dangerous surroundings such as the middle of the sea.

The gas and black crude oil which come out of the ground are carried through pipes to a refinery. The oil is separated into fuels such as petrol, diesel and kerosene. Oil is also used to make plastics, fibres for clothing, paints and medicines.

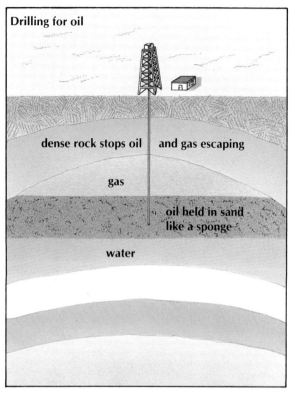

Drilling for oil

dense rock stops oil and gas escaping

gas

oil held in sand like a sponge

water

▼ Liquid gas is used instead of petrol in many cars in Australia. The gas produces less pollution than petrol but it still adds carbon dioxide to the air.

Cars, lorries, trains, ships and aeroplanes all use energy from oil to make them move. Some power stations use oil as a fuel for generating electricity. But the use of oil as a fuel causes air pollution. Your home or school might be heated by oil, or have a piped supply of natural gas for cooking and heating.

Oil is carried around the world in ships. If a ship sinks, or the oil is spilled, it floats on the surface of the sea, killing birds and sea life.

Like coal, oil and natural gas cannot be replaced once they have been used. They are being used so fast that there are only enough supplies left for another 40 years.

# Cars and energy

Ask an adult to help you with a car survey. Find a safe place near your home or school to count passing cars. You will need paper and a pencil.

Look at each car as it passes. How many people are there in the car? Draw two columns on your paper. Tick one column for each car with just one person inside and tick the other for cars with more than one person. Spend 15 minutes watching. Which column has the most ticks in it? Is energy being used wisely?

Transport uses a lot of energy from oil. Oil can be made to last longer if people use it more carefully. Driving at slower speeds reduces the amount of petrol used. Are there any journeys which you make that could be made using less energy by a different form of transport?

There are about three million cars in Mexico City. Car exhaust fumes cause 83% of the air pollution in the city.

# Which cars use the most energy?

The bigger a car's engine, the more fuel it needs. The amount of fuel cars use is measured in kilometres per litre. The higher the number, the less energy a car uses. Which cars are the most energy-hungry? Collect information from people you know who drive cars or look up the figures in a car magazine in your local library.

## What you can do

* Walk, cycle and use public transport wherever possible instead of using a car.

* Design a poster to encourage people to share a car journey when travelling to school, the shops or to work.

* Persuade people to drive at the speed limit to save energy.

* If you know someone who is buying a new car, ask them to think about changing to one with a smaller engine which will use less energy.

# Electrical energy

How many times today have you used something which needs electricity to make it work? Electrical energy is one of the easiest forms of energy to use.

Electricity is produced in power stations which are often a long way from the places where the energy is needed. The electrical energy which the power stations produce is delivered to your home and school through underground cables.

The cables hanging from these pylons are used to carry electricity away from power stations. The journey of the electricity may continue along underground cables before it reaches the buildings where it is needed.

Power stations waste vast amounts of energy. Only one third of the energy produced by burning coal and oil is turned into electrical energy. The rest is lost to the air as heat. Sometimes, if a power station is near a town, district heating schemes can use this waste energy to heat local buildings and to provide the area with a supply of hot water.

A useful way of saving energy is to use other people's waste energy. District heating schemes can use waste energy from oil refineries and local refuse incinerators as well as from power stations. This means that less electricity from power stations is needed, saving fuel and reducing pollution.

In Sweden and Holland, combined heat and power stations are being built. These are much smaller than normal power stations so they can be built in the middle of towns.

Power stations burn huge amounts of coal and oil. This makes them one of the world's biggest causes of acid rain and the greenhouse effect. The waste gases are sent up high chimneys and then carried along by the wind for many miles, often to other countries.

The gases can be cleaned in the chimneys to reduce acid rain. But this cleaning process raises the cost of the electricity. The process cannot remove carbon dioxide, which then adds to the greenhouse effect.

▶ This scientist is monitoring the effects of acid rain on trees in Sweden.

# Electrical energy test

Electric motors turn electrical energy into motion energy. You can watch this process happen. You will need a piece of thin electrical wire about one metre long, a large iron nail, a battery and a magnetic compass.

**1** Make an electromagnet by winding the wire around the nail leaving the two wire ends long enough to join to the battery.

**2** Put a magnetic compass near one end of your electromagnet. Now connect up the battery and watch the needle of the compass. What happens to the needle when you disconnect the battery? Does it matter where you position the compass?

Electric cars use large batteries to drive an electric motor. But at the moment, these batteries cannot store enough energy to power the car for very long. When scientists solve this problem, people will be able to drive quiet, electric cars which do not pollute the air.

▶ This train is powered by an electric motor, which is used to turn electrical energy into movement energy.

When you run, you are using energy more quickly than when you are walking. You need more power. Power is a measure of how fast energy is used. It is measured in watts (W) or kilowatts (kW). A kW equals 1000 watts.

Most electrical appliances have a label to tell you how much power they use. A light bulb might have 100W marked on it, and an electric heater 3kW. When you are using the heater you are using energy much more quickly than when you are using the light bulb.

## Energy use survey

Make a survey of the electrical machines you use at home or at school. Make sure that each machine is unplugged before you look for the label which tells you how much power it uses. Which machines use the most energy? How could you use your results to help you to save energy at home and at school?

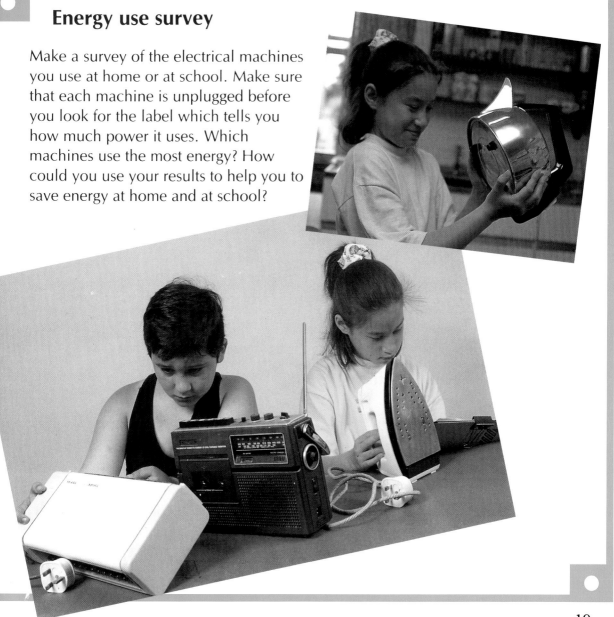

Electricity is sold in units of kilowatt hours. One unit is the amount of energy you would use if you switched on ten 100W light bulbs for one hour (1kW per hour). Ask to see a recent electricity bill to find out how much a unit of electrical energy costs.

If your home is heated by electricity, you could compare the bills for the electricity which was used during last summer and winter. The difference in the bills will give you some idea of how much it costs to heat your home.

Every house which has a supply of electricity has a meter to measure how much energy is used. Ask to see the meter at home or at school. You could keep a daily record of how much energy is used over a week at different times of the year.

Buildings can be insulated to stop heat energy from escaping through the roof, walls and floor. Double-glazed windows also help to keep heat in. Have a look at the bottom of doors at home or at school. Can you think of a way to stop heat energy escaping from here?

## What you can do

By saving electrical energy you will be conserving fuels, reducing pollution and saving money, too.

Persuade people at home to buy low energy bulbs wherever possible.

Switch off lights when you're not using them.

Close doors and don't leave windows open in rooms which are being heated.

If you have central heating at home or at school, ask for the thermostat to be set a few degrees lower. You won't notice the temperature drop and you'll be saving energy.

Ask for a bulb of lower wattage to replace one that you think is too bright.

If there is a washing machine at home, persuade your parents to use a low temperature setting, and then only with a full load.

If you feel cold, put an extra jumper on instead of turning the heating up.

# Nuclear energy

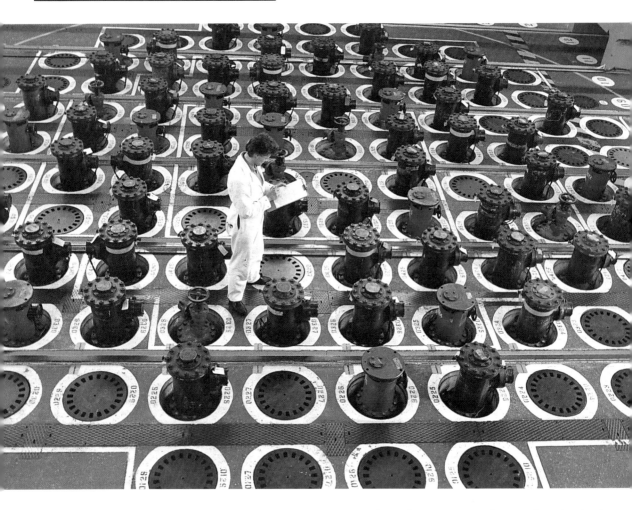

▲ A nuclear power station worker checks the tubes through which uranium fuel rods are loaded into the nuclear reactor.

Nuclear energy is used in some power stations to produce electricity. A substance called uranium is used in this process. One kilogram of uranium produces 20,000 times more energy than burning the same amount of coal or oil.

Everything is made up of tiny particles called atoms. These atoms are so small that billions of them would be needed to make a single grain of sand. Inside each atom is a nucleus, which is held together by nuclear energy.

Uranium atoms are radioactive. This means that their nuclei can be broken apart easily to release nuclear energy. Scientists have learnt to control this process, called fission, in a nuclear reactor.

Many people oppose the use of nuclear energy because of its threat to the environment. Nuclear reactors produce dangerous radioactive waste which would seriously harm or even kill any living thing that came into contact with it.

So nuclear waste has to be buried deep underground. If there are leaks, or if the reactor is damaged by an accident, radioactive waste may escape into the environment.

Nuclear power stations are designed to be used for only 30 years. After that the reactor cannot be demolished for hundreds of years because of the dangerous substances left inside.

Energy is produced when atoms join or fuse together as well as when they are split apart. Scientists are trying to control this fusing process in nuclear reactors.

If they succeed, nuclear fusion may be able to provide us with an unlimited and safe supply of energy for the future, because the process doesn't produce dangerous, radioactive waste.

▼ A scientist collects samples of grass growing outside a nuclear power station to test them for levels of radiation.

# Renewable energy: water

As the Earth's stores of coal, oil and gas run out, people are searching for alternative sources of energy. One source which people have been using for thousands of years is the energy from flowing water.

If you turn on a tap fully open and put your hands underneath you will feel the energy of flowing water. Water wheels make use of the motion energy which is produced by a fast-flowing stream or river.

Water wheels like the one on the right are used in mills to turn the millstone which grinds wheat into flour. As long as the water keeps flowing, it provides plenty of energy without causing pollution.

Water wheels called water turbines are used in hydro-electric power stations to provide much greater amounts of energy.

This dam provides a continuous supply of water to a hydro-electric power station.

A hydro-electric power station uses the energy from water which is stored in a dam. Water flows out of the dam through a pipe to a water turbine which turns a large generator. The electricity produced is carried along cables to the places where it is needed.

◀ Dam-building in Vietnam.

Hydro-electricity cannot be produced without water from a dam, but the building of dams often damages the local environment. The water behind a dam forms a lake which floods large areas of land, destroying wildlife habitats. Sometimes people's lives are disrupted when whole villages have to be moved.

The sea is another source of water energy. If you have ever paddled in the sea you'll have felt the energy of water on your legs. Engineers are trying to find ways of using wave energy. In Norway and France the energy produced by the daily rising and falling of the tides is used to drive electrical generators.

This reservoir extends for 100 km behind a dam in Brazil. The reservoir was created at the cost of millions of trees which were drowned when the dam was built.

25

# Renewable energy: wind

Look out of the window. Is it windy today? How do you know? The wind is an invisible source of energy which can make things move.

Wind surf boards and sailing boats get their movement energy from the power of the wind.

Windmills use wind energy. The wind turns sails which drive shafts and gears inside the windmill to turn millstones or water pumps. What do you think happens when the wind stops?

Some windmills have been built which can use the wind's energy to generate electricity. These wind-powered generators provide energy for small amounts of electricity.

But windmills like these will never replace coal and oil burning power stations. A single power station produces 2000 times more energy than one large wind-powered generator.

Compare the windmill on the left of this photograph with the 'wind farm' of wind-powered generators on the right.

# Renewable energy: sun

Is the sun shining today? Even if the sky is cloudy, the sun is still sending out light and heat energy.

This solar energy can be very harmful if you do not treat it properly. Never look directly at the sun. The light energy is so strong that it can blind you. In the summer, when the sun is shining, you need to take care to avoid sunstroke or sunburn.

Plants need solar energy to make food. People have learned how to use solar energy. Solar panels on the roofs of buildings can be used to trap the sun's heat to warm up water. They work best when it is warm and sunny.

## Testing solar energy

Use solar energy to heat up some water. It is best to do this on a warm, sunny day. You will need three shallow plastic trays, some black and white plastic bags and some aluminium foil.

**1** Cover one tray in black plastic, the other in white plastic and the third in aluminium foil. Make sure that the plastic or foil overlaps the sides of each tray. Put the three trays outside in a sunny position.

**2** Pour water into each tray until it is 1cm deep. Every ten minutes, feel the temperature of the water with your hand. Which material allows the sun to heat up the water the most?

This television runs on energy from the sun.

Solar cells are used to change energy from the sun into electrical energy. Perhaps you've seen a solar-powered calculator. People have even made small cars which are powered by solar energy and there are satellites in space which use large solar cell panels.

Unfortunately, solar cells are not very good at producing large amounts of electrical energy, so they cannot replace power stations.

# Renewable energy: geothermal

Deep down beneath your feet there is heat energy hidden in the underground rocks. Ten kilometres down it can be hot enough to melt the rocks. Sometimes this molten rock finds a way to the surface and erupts as a volcano.

In countries like Iceland, New Zealand, Japan and Britain, this pollution-free geothermal energy is close enough to the surface for people to use. In some areas, for example, hot water springs provide a ready-to-use energy source.

In other places, holes are drilled to reach the rocks, and cold water pumped down to be heated. The hot water which comes back up is piped under the roads and used to heat local buildings.

▲ Sometimes heat energy from inside the Earth reaches the surface of the ground in the form of bubbling, boiling mud.

▼ The water which is used to heat this outdoor swimming pool is produced by the geothermal power station in the background.

# Renewable energy: biomass

Plant and animal waste, or biomass, can be used to produce energy.

In Brazil, sugar cane is used to make alcohol, which can be used instead of petrol to fuel vehicles. Although alcohol does not cause pollution like petrol, it is more expensive to produce. Large areas of land are needed to grow the sugar cane. This leaves less land for people to grow food on.

In China and India, dead plants and animal manure are left to rot in tanks called digesters to produce a gas which is used for cooking. However, the process is not suitable for larger-scale gas production. The digesters would use up more energy than they would actually produce.

A digester

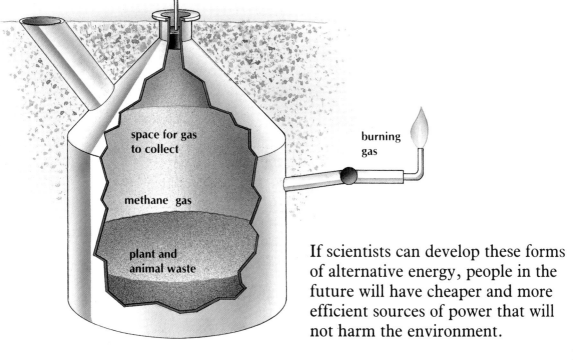

space for gas to collect

methane gas

plant and animal waste

burning gas

If scientists can develop these forms of alternative energy, people in the future will have cheaper and more efficient sources of power that will not harm the environment.

# Useful addresses

If you would like to find out more about the ideas in this book, write to any of these organisations:

**British Coal Opencast**
200 Lichfield Lane, Berry Hill, Mansfield, Nottinghamshire, NG18 4RG.
**Centre for Alternative Technology**
Llwyngwern Quarry, Machynlleth, Powys, SY20 9AZ.
**The Forestry Industry Committee of Great Britain**
Agriculture House, Knightsbridge, London, SW1X 7NJ.
**Friends of the Earth (UK)**
26–28 Underwood Street, London, N1 7JQ.
**Friends of the Earth (Australia)**
Chain Reaction Co-operative, P.O. Box 530E, Melbourne, Victoria 3001.
**Friends of the Earth (New Zealand)**
P.O. Box 39-065, Auckland West.
**Greenpeace (UK)**
30–31 Islington Green, London, N1 8XE.
**Greenpeace (Australia)**
Studio 14, 37 Nicholson Street, Balmain, New South Wales 2041.
**Greenpeace (New Zealand)**
Private Bag, Wellesley Street, Auckland.
**Information Services, British Nuclear Fuels plc**
Risley, Warrington, WA3 6AS.
**UK Atomic Energy Authority Education Service**
Building 354, West, Harwell Laboratory, Didcot, Oxfordshire, OX11 0RA.
**Understanding Electricity**
30 Millbank, London, SW1P 4RD.

# Index